The Purpose of the Environmental Crisis

A Reinterpretation of Hölderlin's Philosophy

Neil Paul Cummins

Cranmore Publications

Copyright © 2011 by Neil Paul Cummins

All rights reserved. This book, or parts thereof, may not be reproduced in any form without permission.

A catalogue record for this book is available from the British Library

ISBN: 978-1-907962-04-2

Published by Cranmore Publications

Reading, England

For Hölderlin

The German Romantic Friedrich Hölderlin developed a unique perspective on the relationship between humankind and the rest of nature. He believed that humanity has a positive role to play in cosmic evolution, and that modernity is the crucial stage in fulfilling this role. In this book I will be arguing for a reinterpretation of his ideas regarding the position of humankind in cosmic evolution, and for an application of these ideas to the 'environmental crisis' of modernity. This reinterpretation is of interest because it entails an inversion of the conventional notion of causality in the 'environmental crisis'; instead of humans 'harming' nature, in the reinterpretation it is nature that causes human suffering.

Contents

Preface 9

Introduction 13

1 Hölderlin's philosophy of human nature, cosmic evolution and modernity 17

2 Interpretations of Hölderlin and his concept of fate 22

3 A Reinterpretation of the Human in Cosmic Evolution 41

4 Objections to the Reinterpretation 63

5 Conclusion 68

Bibliography 72

Preface

Since I first came across the writings of Friedrich Hölderlin I have felt that there is some kind of ineffable connection between us. Perhaps the basis of this connection is a very similar view of the place of humanity in the evolution of the cosmos *combined with* a similarity of character and personality.

Hölderlin formulated his view of the human species and its place in universal evolution at the end of the eighteenth century, long before the onset of the environmental crisis. In Hölderlin's era there were definitely no concerns about human-induced global warming.

Being born in the late twentieth century I have been able to see the human species evolve to a further point than Hölderlin had the luxury of. I have been able to live at a time when there is a realisation that the human species has initiated a planetary environmental crisis which includes human-induced global warming. I have thus been able to think about how the environmental crisis of modernity fits into the bigger picture of cosmic evolution that Hölderlin envisioned in the eighteenth century.

The environmental crisis has always been a key component in my thinking about the place of the human species in universal evolution. I expect that if Hölderlin were alive today that he would agree with me that the environmental crisis has a purpose. I very much hope that Hölderlin would approve of the reinterpretation of his work which is offered in the following pages.

Introduction

Friedrich Hölderlin, one of the German Romantics, developed a distinctive viewpoint on the relationship between humankind and the rest of nature. His ideas are of particular interest because he yearned for an end to human suffering, but was also firmly convinced that humankind was inevitably destined to be separated from nature, and thereby destined to endure suffering. Hölderlin envisioned a positive role for humanity in cosmic evolution, a role which has significant implications for both human nature and cultural evolution. In this book I will be outlining Hölderlin's ideas, and arguing for an application of them to the 'environmental crisis' of modernity. Hölderlin's conception of the human-nature relationship as part of an unfolding process of cosmological change seems to be of great relevance today, an age that is characterized by belief in the meaninglessness of human existence, and by concern about the way that we have altered the pre-human conditions of the Earth. Hölderlin's views provide a unique perspective on modernity that is worthy of serious consideration.

I start by outlining Hölderlin's views on the role of humankind in universal evolution. I then review the secondary literature on Hölderlin that relates to these ideas. I proceed to argue that Hölderlin's philosophy is applicable to, and gives a unique

Introduction

perspective on, the 'environmental crisis' of modernity. I contend that the existing secondary literature on Hölderlin has not recognized this, and that a reinterpretation of the role of humanity in Hölderlin's philosophy of cosmic evolution is therefore required. My central claim is that for Hölderlin, modernity and the related notion of the contemporary 'environmental crisis' is a necessary stage of cosmic evolution, and thus that it is far from a 'crisis'. Rather it is a necessary stage of disharmony that will inevitably be followed by a re-conquered harmony. I will argue that for Hölderlin this disharmony is characterized by the environmental

The Purpose of the Environmental Crisis

changes that are resultant from the development of technology.

Chapter 1

Hölderlin's philosophy of human nature, cosmic evolution and modernity

The starting point of Hölderlin's philosophy is that there must be a basic unknowable reality which precedes self-consciousness wherein subjects and objects are not in existence but are both part of a 'blessed unity of being'. He describes this unity as, "Where subject and object simply are, and not just partially, united...only there and nowhere else can there be talk of being."[1] He argues that the 'blessed unity of being' (which he also refers to as 'nature') is responsible for the coming into existence of humanity through using its power to initiate a division of itself into subjects and objects. This division of being causes the emergence of judgement. Hölderlin states that, "'I am I' is the most fitting example of this concept of judgement...[as] it sets itself in opposition to the *not-I*, not in opposition to *itself*."[2]

The division means that human beings are not capable of actions that are independent of nature; Hölderlin states that, "all the streams of human

---

[1] Friedrich Hölderlin, 'Being Judgement Possibility', in J. M. Bernstein (ed.), *Classic and Romantic German Aesthetics*, Cambridge, Cambridge University Press, 2003, p. 191.
[2] Ibid., p. 192.

activity have their source in nature."[3] It is revealing to compare this claim with the words of Hölderlin's character Hyperion, "What is man? – so I might begin; how does it happen that the world contains such a thing, which ferments like a chaos or moulders like a rotten tree, and never grows to ripeness? How can Nature tolerate this sour grape among her sweet clusters?"[4] For Hölderlin, man is the 'violent' being, whose coming into existence in opposition to the rest of nature was *initiated* by nature.

Hölderlin sees this opposition between man and the rest of nature as culminating in modernity –

---

[3] Alison Stone, 'Irigaray and Hölderlin on the Relation Between Nature and Culture', in *Continental Philosophy Review,* vol. 36, no. 4, 2003, p. 423.
[4] Friedrich Hölderlin , 'Hyperion', in Eric L. Santner (ed.), *Hyperion and Selected Poems,* New York, Continuum, 1990, p. 35.

an era that he claims is characterised by the absence of the gods. In *Brot und Wein* Hölderlin writes, "Though the gods are living, Over our heads they live, up in a different world...Little they seem to care whether we live or do not."[5] A key question for Hölderlin is how we deal with this separation. He envisions two possibilities – the 'Greek' response which is to dissolve the self and die, and the 'Hesperian' response of a living death.

Hölderlin came to view the 'Greek' response as hubristic, it being based on an anthropocentric desire to oppose the division initiated by nature. He thus sees the 'Hesperian' response of living and

---

[5] Ibid., p. 185.

carrying out actions that are dependent on nature for their origination as the appropriate non-hubristic response to our separation. Hölderlin's position is that as nature created the separation, *only* nature can bring the separation to an end. He sees this process of separation and reconnection as part of a broader cosmic picture wherein nature is an unfolding organism rather than a huge mechanism. This organismic view enables him to envision teleological processes in nature which give rise to his claim that there will be, "eternal progress of nature towards perfection."[6]

---

[6] Ronald Peacock, *Hölderlin,* London, Methuen & Co. Ltd, 1938, p. 36.

Chapter 2

Interpretations of Hölderlin and his concept of fate

In this chapter I set out my view of Hölderlin's conception of fate – that all human actions are part of the evolution of nature towards perfection. I do this by reviewing the existing scholarly literature on Hölderlin and showing that whilst these interpretations all recognise parts of Hölderlin's conception of fate that they do not capture the whole of it. I start with interpretations of human nature, move on to cosmic processes, and finally consider the role of modernity within these processes.

Interpretations of Hölderlin

At the level of the human there is a general consensus in the literature that Hölderlin's position is that humans are endowed by nature with qualities that shape human nature, and that this inevitably shapes human interactions with the rest of nature. There are various names in the literature for the qualities which are endowed to humans. Dennis J. Schmidt refers to the qualities present in humans as their 'formative drive.' He claims that, "Hölderlin suggests that human nature and practices are to be understood by reference to a formative drive which expresses itself as a constant need for 'art'."[7] In a similar vein, Thomas Pfau argues for an 'intellectual

---

[7] Dennis J. Schmidt, *On Germans and Other Greeks,* Indiana University Press, 2001, p. 139.

intuition.' He states that, "Hölderlin recasts the convergence of "freedom and necessity" as the most primordial synthesis of intellect and intuition itself, a synthesis which takes place within the subject itself. He thus approaches what Kant had repeatedly ruled out as an "intellectual intuition"."[8]

In agreement with Schmidt and Pfau, Franz Gabriel Nauen argues that for Hölderlin, "all men do in fact have the same basic character…all human activity can be derived from the same *elemental drive* in human nature."[9] The 'formative drive' / 'intellectual intuition' / 'elemental drive' identified

---

[8] Thomas Pfau, *Friedrich Hölderlin: Essays and Letters on Theory*, New York, SUNY Press, 1988, p. 15.

[9] Franz Gabriel Nauen, *Revolution, Idealism and Human Freedom: Schelling, Hölderlin and Hegel and the Crisis of Early German Idealism,* Indiana University Press, 2001, p. 139.

in the literature explains why man can be seen as the 'violent' being. Human nature is to engage in 'art', to utilize the resources of nature so that culture can be generated and sustained. This generation of human culture actually benefits nature as a whole, but it requires large-scale modification of parts of non-human nature. The destiny of man is thus a disruptive one. It is clear that it is also an undesirable one. Nauen states that for Hölderlin, "Even war and economic enterprise serve to fulfil the destiny of man, which is to "multiply, propel, distinguish and mix together the life of Nature"."[10]

So Hölderlin sees human nature, economic production and even war as parts of a broader

---

[10] Ibid.

cosmic evolutionary process; the universe *as a whole* is seen as evolving to perfection. There will inevitably be aspects of this evolution that from a narrow perspective could be viewed as 'less than perfect'. These negative aspects of the evolutionary process – from war, to the presence of evil in its entirety – have to be seen as inescapable parts of the whole process.

The key point is that for Hölderlin the cosmic evolutionary process *ends* in perfection. Thus, Ronald Peacock argues that, "the division produced by conflict is followed by a re-conquered harmony."[11] Similarly, Anselm Haverkamp argues that an interpretation of the poems *Andenken* and *Mnemo-*

---

[11] Peacock, *Hölderlin,* p. 22.

syne is the expression, 'where danger threatens, salvation also grows.'[12] Whilst, Martin Heidegger translates the opening lines of *Patmos* as, "But where danger is, grows the saving power also."[13] Hölderlin's view is clearly that from a narrow and short-term perspective danger and conflict are often the norm, but that these things actually play a part in bringing about a greater harmony in the future. In the long-term they are all part of the evolution of the whole universe to perfection.

Cosmic evolution is thus one long process of disharmonies and inevitably following harmonies.

---

[12] Anselm Haverkamp, *Leaves of Mourning: Hölderlin's Late Work,* New York, SUNY Press, 1996, p. 48.

[13] Martin Heidegger, 'The Question Concerning Technology', in R.C. Scharff and V. Dusek (eds.), *Philosophy of Technology: The Technological Condition – An Anthology,* Oxford, Blackwell Publishing, 2003, p. 261.

The Purpose of the Environmental Crisis

Peacock argues that Hölderlin's vision is of a, "harmonised process of life which comprises within itself the rhythmic movement from chaos to form and back again, and an emotional experience of this which in the sphere of nature knows only the one rapture, but in the human sphere suffering and joy."[14] It is revealing that this interpretation sees 'violent' humans as suffering, whilst nature is purely rapturous. This clearly sheds light on the question posed by Hölderlin's character Hyperion: "How can Nature tolerate this sour grape among her sweet clusters?"[15] The answer seems to be that human

---

[14] Peacock, *Hölderlin*, p. 22.
[15] Hölderlin, 'Hyperion', p. 35.

'violence' *enables* nature to be rapturous. As part of this rapture humans experience suffering.

Why should suffering be a uniquely human experience? To explain this Peacock cites part of a letter from Hölderlin to his brother, "Why can they [humans] not live contented like the beasts of the field? he asks: and replies that this would be as unnatural in man, as in animals the tricks, or arts, man trains them to perform. Thus he establishes that the arts of man are natural to man. Culture, then, derives from nature; and the impulse to it is the characteristic which distinguishes man from the rest of creation."[16]

---

[16] Peacock, *Hölderlin,* p. 36.

The human impulse to culture has culminated in the era of modernity. Hölderlin sees this period as one of great significance as he sees it as a historical epoch that is characterised by the *absence of the gods*. To be consistent with his views on harmonised evolution to perfection there must be a reason for this absence. Indeed, Peacock argues that Hölderlin thinks that, "a godless age is part of a divine mystery, it is as necessary as day, ordained by a higher power."[17] Furthermore, Heidegger claims that the gods are still present, despite their absence: "man who, even with his most exulted thought could hardly penetrate to their Being, even though, with

---

[17] Ibid., p. 92.

Interpretations of Hölderlin

the same grandeur as at all time, they were somehow there."[18]

The absence of the gods in modernity is deeply related to the contemporary danger that exists in modernity. It should be remembered that this danger cannot be a cause for concern for Hölderlin – as all dangers are inevitably followed by regained harmonies. Nevertheless, Heidegger attempts to identify the exact danger that Hölderlin believed is present in modernity. Heidegger claims that, "the essence of technology, enframing, is the extreme danger."[19] It must follow that for Heidegger, "precisely the essence of technology must harbor in itself

---

[18] Martin Heidegger, *Existence and Being*, London, Vision Press Ltd., 1956, p.190.
[19] Heidegger, 'The Question Concerning Technology', p. 261.

the growth of the saving power."[20] He sees this as occurring when the essential unfolding of technology gives rise to the possibility of opening up a "free relation" with technology which is inclusive of non-instrumental possibilities.[21]

In an interpretation of the 1802 hymn *Friedensfeier*, Richard Unger draws out Hölderlin's views on the absence of the gods in modernity.[22] In *Friedensfeier* the entire span of Western civilization is characterised as a thunderstorm which is ruled by a "law of destiny" which ensures that a certain

---

[20] Ibid.
[21] R.C. Scharff and V. Dusek, 'Introduction to Heidegger on Technology', in R.C. Scharff and V. Dusek (eds.), *Philosophy of Technology: The Technological Condition – An Anthology*, Oxford, Blackwell Publishing, 2003, p. 248.
[22] Richard Unger, *Friedrich Hölderlin*, Boston, Twayne Publishers, 1984, pp. 100-105.

Interpretations of Hölderlin

amount of "work" is achieved. Unger argues that it is clear that this "work", "is the product of the storm itself and that it designates the harmonious totality of earthly existence during the coming era."[23] The end of the "storm" of modernity enables the arrival of a mysterious "prince" who makes it possible that, "men can now for the first time hear the "work" that has been long in preparation "from morning until evening"."[24]

Following the inevitable successful accomplishment of the "work" of Western civilization, the great Spirit will disclose a Time-Image which will, "be a comprehensive depiction of the historical

---

[23] Ibid., p. 102.
[24] Ibid., p. 101.

process and its triumphant result."[25] Unger argues that, "the Image shows that there is an alliance between the Spirit of history and the elemental divine presences of nature – for the natural elements with which man has always worked have played integral and essential parts in man's history."[26] The triumphant result of the actions of humankind in modernity is clearly an example of a re-conquered harmony that follows division.

In Unger's interpretation of *Friedensfeier* we have a picture of modernity in which humans are carrying out "work" under a "law of destiny". The crucial factor is that humanity is ignorant that it is

---

[25] Ibid., p. 104.
[26] Ibid., p. 105.

Interpretations of Hölderlin

working under a "law of destiny" in modernity, until modernity has ended. It is then that through the Time-Image the great Spirit reveals the successful outcome of modernity, and the *nature and value* of the accomplished "work". This is a prime example of a short-term and narrow perspective entailing the perception of a lack of destiny and of needless suffering, whilst in the longer-term the same events are seen to be an inevitable part of a broader positive outcome – the evolution of the universe to perfection.

This difference of perspectives can explain an apparent contradiction in the literature between Unger's interpretation of *Friedensfeier,* and Schmidt's analysis of Hölderlin's 1801 letter to

The Purpose of the Environmental Crisis

Bohlendorff. This letter was written only one year before *Friedensfeier* and Schmidt claims that in it Hölderlin's position is, "that the peculiar flow of modernity is the lack of destiny."[27] The apparently contradictory views of Unger and Schmidt can be reconciled through recalling Peacock's interpretation that, "a godless age is part of a divine mystery, it is as necessary as day, ordained by a higher power,"[28] and comparing it to Unger's claim that men are blind to the point of the "work" that they have been carrying out until the "storm" of Western civilization has passed.

---

[27] Schmidt, *On Germans and Other Greeks,* p. 137.
[28] Peacock, *Hölderlin,* p. 92.

The comparison reveals that the "law of destiny" applies to the activities of *humanity as a collective* in Western history, activities that are ordained by a higher power for a specific purpose. In contrast, the "lack of destiny" applies to *individual human beings*. This difference arises because individual humans are unaware that their actions are part of an inevitably unfolding cosmic plan, it is only the fruition of the plan than enables realization. Instead, humans believe that they have free will and live in a meaningless age. Therefore, modernity can at one and the same time be characterized as both a period governed by a "law of destiny" and a period constituted by a "lack of destiny". The difference is purely one of perspective.

The Purpose of the Environmental Crisis

This conception of modernity as simultaneously being a period of a "lack of destiny" and a "law of destiny" raises the issue of anthropocentricism. If human attitudes and actions towards nature are in the interests of nature, then it seems that there is no such thing as a *truly* anthropocentric attitude. The appropriate attitude that humans should take to the objective side of nature, given Hölderlin's philosophy, has been addressed by Alison Stone. She argues that because, "according to Hölderlin's thinking, we have become separated from nature by *its* power alone, so it is not within *our* power to undo separation."[29] Therefore, "the appropriately modest

---

[29] Stone, 'Irigaray and Hölderlin on the Relation Between Nature and Culture', p. 424.

Interpretations of Hölderlin

response is to endure separation – to wait, patiently, until nature may change its mode of being."[30] This means that a truly non-anthropocentric environmental view of the rest of nature requires, "the *acceptance* of disenchantment, of separation, of meaninglessness."[31]

This view is concordant with the "lack of destiny" perspective. However, when the "law of destiny" is taken into account, then the hidden meaning is revealed. Furthermore, the whole notion of the attitudes of individual humans then becomes irrelevant. It seems that there cannot be such a thing

---

[30] Ibid.

[31] Alison Stone, *Nature in Continental Philosophy – Week 4, Section V, Friedrich Hölderlin,* [online], http://www.lancaster.ac.uk/depts/philosophy/awaymave/408new/wk4.htm, [accessed 25 October 2005].

as a *truly* anthropocentric attitude, because all attitudes originate from nature, and they all lead to actions which fulfil the "law of destiny". It may seem that our attitudes to nature are of importance, but this is because we believe in a "lack of destiny", and are inevitably blind to the bigger picture of the "law of destiny". Whatever our attitudes as individuals, our relationship with the rest of nature as a collective would be 'for the best'.

Chapter 3

A Reinterpretation of the Human in Cosmic Evolution

The interpretations of Hölderlin that I have reviewed all give an accurate representation of Hölderlin's views. However, they are all partial views. They all miss the 'big picture' of what Hölderlin's views imply about what it means to be a human in the context of cosmic evolution, and the consequent implications for the perspective from which we should view modernity and the 'environmental crisis'. In an attempt to fully grasp these implications I am going to defend the thesis that:

The Purpose of the Environmental Crisis

Hölderlin's philosophy leads to the conclusion that the 'environmental crisis' is a necessary stage in the purposeful evolution of nature towards perfection. This is an interesting thesis because, if accepted, it would supplant the conception of the meaninglessness of human existence with a conception of positive cosmic purpose.

The argument I will be making centres on three key aspects of Hölderlin's philosophy. Firstly, that he believes that nature is purposefully evolving towards perfection. Secondly, that he believes that the achievement of this perfection requires human actions. Thirdly, that he believes that human actions are determined by nature. Acceptance of these three claims leads to the conclusion that human actions are determined by nature as a necessary stage in the purposeful evolution of nature towards perfection. As the 'environmental crisis' of modernity is purely resultant from human actions, a second conclusion inevitably follows. This is that the 'environmental crisis' itself is determined by nature as a necessary stage in the purposeful evolution of nature towards perfection.

I will now present evidence to support the three key claims. The first claim is that Hölderlin's belief is that *nature is purposefully evolving towards perfection*. The universe can either be viewed as a giant mechanism or as an unfolding organism; Hölderlin clearly held the latter view. This conception of the universe explains his belief that nature unfolds in a way that serves its own purposes; that disharmonies are followed by regained harmonies. This is why Peacock claims that Hölderlin believed in, "the eternal progress of nature towards perfection,"[32] and, "the emergence of perfection in the course of natural development."[33]

This firm belief clashed with Hölderlin's personal yearning for immediate perfection in life. His immense desire to see a morally just world was completely at odds with his philosophical belief that the perfection he sought could only be achieved in the course of natural development. The movement to perfection envisioned by Hölderlin is thus a

---

[32] Peacock, *Hölderlin*, p. 36.
[33] Ibid., p. 105.

fatalistic one, an inevitable evolutionary progression towards perfection. Peacock captures this with his claim that for Hölderlin there is an, "acute sense of 'Fate', of inevitability, expressed again and again in his work. Fate is revealed in the process of history… it is inherent in the passage of form to chaos, and of disintegration to a new harmony."[34]

This first claim is the most straightforward of the three. The second claim is that *Hölderlin believes that the achievement of perfection requires human actions*. The starting point in defending this claim is Hölderlin's central belief that nature *used its power* to divide itself and thereby create humankind. This division means that the split was part of

---

[34] Ibid., p. 93.

The Purpose of the Environmental Crisis

the evolutionary process rather than a random occurrence. We can ask ourselves why this may have been a necessary occurrence. An initial answer seems to be Nauen's claim that, "Even war and economic enterprise serve to fulfil the destiny of man, which is to "multiply, propel, distinguish and mix together the life of Nature"."[35]

In *The Perspective from which we Have to look at Antiquity* Hölderlin asserts that, "antiquity appears altogether opposed to our primordeal drive which is bent on forming the unformed, to perfect the primordial-natural so that man, who is born for art, will naturally take to what is raw, uneducated,

---

[35] Nauen, *Revolution, Idealism and Human Freedom: Schelling, Hölderlin and Hegel and the Crisis of Early German Idealism*, p. 139.

A Reinterpretation of the Human

childlike rather than to a formed material where there has already been pre-formed [what] he wishes to form."[36] In a letter to his brother he also asserts that, "the impulse to art and culture...is really a service that men render nature."[37]

The source of Hölderlin's primordeal drive to art is nature, because it is nature that created us and endowed us with our capabilities. This is clear from Peacock's interpretation that, "Man cannot be master of nature; his arts, *necessary though they may be in the scheme of things,* cannot produce the substance which they mould and transform; they

---

[36] Friedrich Hölderlin, 'The Perspective from which We Have to Look at Antiquity', in Thomas Pfau (ed.), *Friedrich Hölderlin: Essays and Letters on Theory,* New York, SUNY Press, 1988, p. 39.
[37] Peacock, *Hölderlin,* p. 37.

can only develop the creative force, which in itself is eternal and not their work."[38]

Hölderlin's primordeal drive to art in humans has inevitably led to the epoch of modernity. Human actions in this epoch appear to be central to the achievement of perfection. Hölderlin claims that modernity is an epoch that, "is as necessary as day, ordained by a higher power."[39] Furthermore, humans have been involved in "work" in modernity that is clearly constitutive of the importance of the epoch. This is clear from Unger's interpretation of *Friedensfeier* in which the "law of destiny" ensures that a certain amount of human "work" is done. The

---
[38] Ibid.
[39] Ibid., p. 92.

crucial factor is that humanity is ignorant that it is working under a "law of destiny" in modernity, until modernity has ended. It is then that through the Time-Image the great Spirit reveals the successful outcome of modernity, and the nature and value of the accomplished "work".

There is no doubt that in Hölderlin's view human actions and their resultant "work" in modernity are part of purposeful evolution to perfection. What is interesting is the exact nature of the "work". There is an obvious connection between the "work" of modernity (*Friedensfeier*) and the "danger" we face in modernity (*Patmos*). Heidegger's interpretation of *Patmos* that, "the essence of technology, enframing, is the extreme danger,"[40] makes it clear that the "work" of modernity is the development of technology. In fact, technological development in modernity seems to be the culmination of Hölderlin's primordeal drive to art. Furthermore, it is very hard to think of any other distinctive aspects of modernity that are resultant from human actions, present an extreme danger, and have cosmic significance. Therefore, for Hölderlin, the achievement of perfection seems to require the human development of technology.

It is interesting that Heidegger sees the danger we face from the "work" of modernity as the essence of technology rather than actual technology. Andrew

---

[40] Heidegger, 'The Question Concerning Technology', p. 261.

A Reinterpretation of the Human

Feenberg has criticised Heidegger for this abstract concentration on essences rather than the actual technology itself.[41] A "Feenberg interpretation" of *Patmos* seems to be more in accordance with Hölderlin's views than the "Heidegger interpretation", as Hölderlin's philosophy is grounded in actualities rather than essences. Hölderlin sees a positive role for actual technology in cosmic evolution; this means that *actual technology* has a cosmic purpose. Therefore, it seems that both the danger we face, and the saviour, must be the *actual* technology developed by human actions.

---

[41] Andrew Feenberg, 'Critical Evaluation of Heidegger and Borgmann', in R.C. Scharff and V. Dusek (eds.), *Philosophy of Technology: The Technological Condition – An Anthology*, Oxford, Blackwell Publishing, 2003, pp. 327-337.

The Purpose of the Environmental Crisis

The importance of the human split from the rest of nature can also be seen in the words of Hölderlin's character *Hyperion:* "How should I escape from the union that binds all things together? We part only to be more intimately one, more divinely at peace with all, with each other. We die that we may live."[42] Human actions are thus depicted as a 'living death' that is necessary for the life (and continued movement to perfection) of nature as a whole. This explains Peacock's interpretation that, "the sphere of nature knows only the one rapture, but in the human sphere [there is] suffering and joy."[43]

---

[42] Hölderlin, 'Hyperion', p. 123.
[43] Peacock, *Hölderlin,* p. 22.

A Reinterpretation of the Human

The third claim is that *Hölderlin believes that human actions are determined by nature.* There are many passages in Hölderlin's novel *Hyperion* that attribute the responsibilities for human actions to a power or god: "There is a god in us who guides destiny as if it were a river of water, and all things are his element."[44]....."oh forgive me, when I am compelled! I do not choose; I do not reflect. There is a power in me, and I know not if it is myself that drives me to this step."[45]....."I once saw a child put out its hand to catch the moonlight; but the light went calmly on its way. So do we stand trying to hold back everchanging Fate. Oh, that it were possible but

---

[44] Hölderlin, 'Hyperion', p. 11.
[45] Ibid., p. 79.

The Purpose of the Environmental Crisis

to watch it as peacefully and meditatively as we do the circling stars."[46]....."Man can change nothing and the light of life comes and departs as it will."[47]....."We speak of our hearts, of our plans, as if they were ours; yet there is a power outside of us that tosses us here and there as it pleases until it lays us in the grave, and of which we know not where it comes nor where it is bound."[48]

Hölderlin's belief in the lack of human free will is perhaps clearest in his claim in a letter to his mother regarding the views of Spinoza that, "one *must* arrive at his ideas if one wants to explain

---

[46] Ibid., p. 22.
[47] Ibid., p. 127.
[48] Ibid., p. 29.

A Reinterpretation of the Human

everything."[49] Spinoza's ideas can be summed up as, "Nature in all its aspects is governed by necessary laws, and human being no less than the rest of nature is determined in all its actions and passions, contrary to those who conceive of it as 'a dominion within a dominion'."[50]

In order to make abundantly clear Spinoza's - and thus Hölderlin's – views on a lack of human free will here are two quotes from Spinoza: "I say that thing is free which exists and acts solely from the necessity of its own nature...I do not place Freedom

---

[49] Friedrich Hölderlin, 'No.41: To his Mother', in Thomas Pfau (ed.), *Friedrich Hölderlin: Essays and Letters on Theory,* New York, SUNY Press, 1988, p. 120.
[50] Moira Gatens, *Imaginary Bodies: Ethics, Power and Corporeality,* London, Routledge, 1996, p. 111.

The Purpose of the Environmental Crisis

in free decision, but in free necessity."[51] And, "a stone receives from an external cause, which impels it, a certain quantity of motion, with which it will afterwards necessarily continue to move...Next, conceive, if you please, that the stone while it continues in motion thinks, and knows that it is striving as much as possible to continue in motion. Surely this stone, inasmuch as it is conscious only of its own effort, and is far from indifferent, will believe that it is completely free, and that it continues in motion for no other reason than because it wants to. And such is the human freedom which all men boast that they possess, and which consists solely in this,

---

[51] Benedict de Spinoza, 'LVIII: To Schuller', trans. A. Wolf (ed.), *The Correspondence of Spinoza*, 2nd ed., London, Frank Cass & Co. Ltd., 1966, pp. 294-5.

A Reinterpretation of the Human

that men are conscious of their desire, and ignorant of the causes by which they are determined."[52]

Furthermore, in an interpretation of Hölderlin's *Stutgard,* Peacock argues that, "the laws of growth govern the culture as well as the lives of men...the one process comprehends all things and the one rhythm manifests itself again and again...in the progress of history; in the spiritual life of individuals."[53] In this vision not only human nature, but also the evolution of culture, is seen as an inevitable historical progression. Peacock's interpretation of Hölderlin is that, "man's spirit is but part of the One

---

[52] Ibid., p. 295.
[53] Peacock, *Hölderlin,* p. 25.

Spirit,"[54] which Hölderlin insists is involved in a "movement...through successive historical generations."[55] The spirit of man is thus governed by the larger Spirit of nature. This is the sense in which, "all the streams of human activity have their source in nature."[56]

The nature of the relationship between man's spirit and the Spirit of nature is made clear in the following quote from Hölderlin's character Diotima: "a *unique destiny* bore you away to solitude of spirit as waters are borne to mountain peaks."[57] This concept of individual humans having a unique

---

[54] Ibid., p. 90.
[55] Ibid., p. 114.
[56] Stone, 'Irigaray and Hölderlin on the Relation Between Nature and Culture', p. 423.
[57] Hölderlin, 'Hyperion', p. 122.

A Reinterpretation of the Human

destiny was the view of Johann Herder, who was one of Hölderlin's inspirations. Herder saw nature as a great current of sympathy running through all things which manifested itself in unique inner impulses within different individuals. This means that every human has a unique calling – an original path which they ought to tread. As Herder states, "Each human being has his own measure, as it were an accord peculiar to him of all his feelings to each other."[58] Clearly, for both Herder and Hölderlin, human actions at any one time are determined in accordance with the movements of the One Spirit of nature.

---

[58] Charles Taylor, *Sources of the Self: The Making of the Modern Identity,* Massachusetts, Harvard University Press, 1994, p. 375.

The Purpose of the Environmental Crisis

I have presented evidence for the claims that for Hölderlin: *nature is purposefully evolving towards perfection, the achievement of this perfection requires human actions, and human actions are determined by nature.* Acceptance of these three claims leads to the conclusion that human actions are determined by nature as a necessary stage in the purposeful evolution of nature towards perfection. I now briefly argue that the 'environmental crisis' of modernity is purely resultant from human actions.

The definition of an environmental problem is: "any change of state in the physical environment which is brought about by human interference with the physical environment, and has effects which society deems unacceptable in the light of its shared norms."[59] This definition encapsulates a sliding scale of environmental problems from those that are local and temporary on the one hand, to those that are global and long-lasting on the other. The 'environmental crisis' as a concept has arisen because of the emergence in the last 100 years of an increasing number of environmental problems that are towards the global and long-lasting end of the scale. The 'environmental crisis' is thus purely resultant from the *human actions* which have created environmental problems that are characterised by their global reach and long-lasting nature.

---

[59] Peter B. Sloep and Maris C.E. van Dam-Mieras, 'Science on Environmental Problems', in P. Glasbergen and A. Blowers (eds.) *Environmental Policy in an International Context: Perspectives*, Oxford, Butterworth-Heinmann, 2003, p. 42.

The Purpose of the Environmental Crisis

This means that the above conclusion, that human actions are determined by nature as a necessary stage in the purposeful evolution of nature towards perfection, needs amending. As the 'environmental crisis' is purely resultant from human actions, it too must be part of this purposeful evolution. Therefore, the new conclusion that inevitably follows is: *the 'environmental crisis' is determined by nature as a necessary stage in the purposeful evolution of nature towards perfection.*

Chapter 4

Objections to the Reinterpretation

It could be objected that there are many references to human freedom in Hölderlin's work that would seem to cast doubt on the third claim. This is particularly noticeable in his novel Hyperion. For example, Hyperion states that, "without freedom all is dead."[60] However, this objection is easily answered because these references all appear in Hölderlin's early work, and even then they are more than counterbalanced by the opposing fatalistic

---

[60] Hölderlin, 'Hyperion', p. 117.

views that I have outlined. In his early period Hölderlin was struggling to come to terms with the conflict between his keen moral aspirations for social change on the one hand, and his belief in perfection only arising through natural development on the other. In his later work, as is clear in his endorsement of the 'Hesperian' response to our condition, he firmly accepts the powers of natural development and the determination of human actions by nature. He realizes the futility of pursuing his idealistic moral aspirations because he accepts the illusory nature of human free will.

A further objection could be made that this re-interpretation is pointless because Darwin's theory of evolution, which emerged shortly after Hölderlin's time, gives a view of evolutionary processes that is incompatible with Hölderlin's view that there was a 'blessed unity of being' prior to the arrival of humans. We now know that the emergence of the human species – and its primordeal drive to art – was preceded by four billion years of evolution of life on Earth. It can thus be argued that there was not a 'blessed unity of being' prior to the evolution of humankind.

This is exemplified by the claim of Hans Jonas that the subject-object divide opened up four billion years ago, when, "living substance, by some original act of segregation, has taken itself out of the general integration of things in the physical context, set itself over against the world, and introduced the tension of "to be or not to be" into the neutral assuredness of existence."[61] This certainly does not appear to be a pre-human 'blessed unity of being'. However, it is interesting that Jonas also sees humans as, "a 'coming to itself' of original substance."[62]

It is clear that this Darwinian based objection does not invalidate the views of Hölderlin, or the reinterpretation of them presented in this book. In fact, not only does evolutionary theory perfectly complement Hölderlin's philosophy, his philosophy *needs* it. The idea that nature could use its power to instantaneously create a being as complex as a

---

[61] Hans Jonas, *The Phenomenon of Life: Toward a Philosophical Biology*, Illinois, Northwestern University Press, 2001, p. 4.
[62] Ibid., p. xv.

Objections to the Reinterpretation

human out of the 'blessed unity of being' is hardly defensible. In the light of our knowledge today we can simply reinterpret Hölderlin as claiming that nature used its power four billion years ago to divide the 'blessed unity of being' and create a subject/object divide. As he sees nature as an unfolding and evolving organism, the divide would give rise to human subjects after a sufficient period of time. This, ""coming to itself" of original substance", as Jonas describes it, has in actuality taken approximately four billion years.

Chapter 5

Conclusion

I have claimed that the existing secondary literature has not grasped the full implications of Hölderlin's thought for what it means to be a human in modernity. By drawing together Hölderlin's ideas I have sought to understand his notion of the purpose of human actions, and what this purpose means for the 'environmental crisis'.

Hölderlin's conception of nature is an organism unfolding to perfection. I have argued that he sees modernity as an important stage of this unfolding, which is characterized by the development of

Conclusion

technology through human actions. I have further argued that this means that the 'environmental crisis' of modernity – a side-effect of the development of technology – is also an inevitable stage of this unfolding; it is in the interests of nature. As nature continues to unfold, the disharmony of modernity will be succeeded by a re-conquered harmony. I have argued that Hölderlin's 'saving power' is actual technology, as this seems most consistent with his thought. Heidegger's view, that the 'saving power' is the essencing of technology, seems inconsistent with the positive role of technology in cosmic evolution that is envisioned by Hölderlin.

The Purpose of the Environmental Crisis

The reinterpretation I have outlined clearly entails an inversion of the conventional notion of causality in the 'environmental crisis' of modernity. Humanity is conventionally pictured as harming nature. My thesis has shown that for Hölderlin it is nature that is 'harming' humanity. We have been cast aside out of the rapture of nature into a realm of suffering and self-consciousness, with the purpose of developing technology to serve the purposes of the unfolding nature of which we are a part.

We are left with the question of what our attitudes to nature should be, given this reinterpretation of what it means to be a human in cosmic evolution. The answer is simple. As nature is the source of our individual attitudes, our attitudes to nature must be

Conclusion

in the interests of nature. Our attitudes, whether they are techno-centric, environmentalist, quietist, or nature-exploitative are all correct for us as individuals, because in the aggregate they fulfil the purpose of nature as a whole.

Bibliography

Feenberg, Andrew, 'Critical Evaluation of Heidegger and Borgmann', in R.C. Scharff and V. Dusek (eds.), *Philosophy of Technology: The Technological Condition – An Anthology*, Oxford, Blackwell Publishing, 2003.

Gatens, Moira, *Imaginary Bodies: Ethics, Power and Corporeality*, London, Routledge, 1996.
Haverkamp, Anselm, *Leaves of Mourning: Hölderlin's Late Work*, New York, SUNY Press, 1996.

Heidegger, Martin, *Existence and Being*, London, Vision Press Ltd., 1956.

Heidegger, Martin, 'The Question Concerning Technology', in R.C. Scharff and V. Dusek (eds.), *Philosophy of Technology: The Techno-*

logical Condition – An Anthology, Oxford, Blackwell Publishing, 2003.

Hölderlin, Friedrich, 'Being Judgement Possibility', in J. M. Bernstein (ed.), *Classic and Romantic German Aesthetics,* Cambridge, Cambridge University Press, 2003.

Hölderlin, Friedrich, 'Hyperion', in Eric L. Santner (ed.), *Hyperion and Selected Poems,* New York, Continuum, 1990.

Hölderlin, Friedrich, 'No.41: To his Mother', in Thomas Pfau (ed.), *Friedrich Hölderlin: Essays and Letters on Theory,* New York, SUNY Press, 1988.

Hölderlin, Friedrich, 'The Perspective from which We Have to Look at Antiquity', in Thomas Pfau (ed.), *Friedrich Hölderlin: Essays and Letters on Theory,* New York, SUNY Press, 1988.

Jonas, Hans, *The Phenomenon of Life: Toward a Philosophical Biology,* Illinois, Northwestern University Press, 2001.

Nauen, Franz Gabriel, *Revolution, Idealism and Human Freedom: Schelling, Hölderlin and Hegel and the Crisis of Early German Idealism,* Indiana University Press, 2001.

Peacock, Ronald, *Hölderlin,* London, Methuen & Co. Ltd, 1938.

Pfau, Thomas, *Friedrich Hölderlin: Essays and Letters on Theory,* New York, SUNY Press, 1988.

Scharff, R. C., and Dusek, V., 'Introduction to Heidegger on Technology', in R.C. Scharff and V. Dusek (eds.), *Philosophy of Technology: The Technological Condition – An Anthology,* Oxford, Blackwell Publishing, 2003.

Schmidt, Dennis J., *On Germans and Other Greeks,* Indiana University Press, 2001.

Sloep, Peter B., and Dam-Mieras, Maris C.E. van, 'Science on Environmental Problems', in P. Glasbergen and A. Blowers (eds.) *Environmental Policy in an International Context: Perspectives,* Oxford, Butterworth-Heinmann, 2003.

Spinoza, Benedict de, 'LVIII: To Schuller', trans. A. Wolf (ed.), *The Correspondence of Spinoza,* 2nd ed., London, Frank Cass & Co. Ltd., 1966.

Stone, Alison, 'Irigaray and Hölderlin on the Relation Between Nature and Culture', in *Continental Philosophy Review,* vol. 36, no. 4, 2003. Stone, Alison, *Nature in Continental Philosophy – Week 4, Section V, Friedrich Hölderlin,*[online],
http://www.lancaster.ac.uk/depts/philosophy/awaymave/408new/wk4.htm
[accessed 25 October 2005].

Taylor, Charles, *Sources of the Self: The Making of the Modern Identity,* Massachusetts, Harvard University Press, 1994.

The Purpose of the Environmental Crisis

Unger, Richard, *Friedrich Hölderlin*, Boston, Twayne Publishers, 1984.

Other books by the author:

Is the Human Species Special? : Why human-induced global warming could be in the interests of life

Should I be a Vegetarian? : A personal reflection on meat-eating, vegetarianism and veganism

How Much of Man is Natural? : Two versions of the international prize winning essay

www.ingramcontent.com/pod-product-compliance
Lightning Source LLC
Chambersburg PA
CBHW071410040426
42444CB00009B/2175